Anatomy of a Lab Accident

Lessons We Should Have Learned, But Didn't

by
Chip Albright
&
A. Kaylene Ray

FHC FUME HOOD CERTIFIED

A Fume Hood Certified Publication
Fume Hood Certified, LLC – Phoenix, AZ USA

Disclaimer

This publication is designed to provide accurate and authoritative information with regard to the subject matter covered. It is sold with the understanding that the contents do not render professional advice. If you require professional assistance or expert advice, you should seek the services of a licensed professional.

Any company names, logos, or images used that are trademarked or copyrighted by others are acknowledged to be the property of the trademark or copyright owner.

This book was created with assistance from artificial intelligence resources. The content has been reviewed and edited by a human. For more information on the extent and nature of AI usage please contact Chip Albright.

Paperback: ISBN ISBN 978-1-7357110-5-8

Ebook: ISBN 978-1-7357110-2-7

First paperback edition July 2025.

Printed in the USA

Published By:

Fume Hood Certified, LLC

PO Box 71477

Phoenix, AZ 85850 USA

www.fumehoodcertified.com

experts@fumehoodcertified.com

+1 623-662-0907

Table of Contents

About the Authors

Chip Albright

This book is the product of Chip Albright's 40-year journey through the world of laboratory safety and fume hood innovation. Chip began his career in 1985 at the Kewaunee Scientific fume hood factory in Lockhart, Texas, where he helped design his first hood—an experience that sparked a lifelong commitment to safer lab environments. Over the decades, he held leadership roles at multiple industry pioneers, including Jamestown-JMP and Collegedale LLC, where he developed several generations of performance-driven fume hoods, including the Isolator and the G3 series.

As standards evolved—from the introduction of ASHRAE 110 in 1985 to the growing influence of OSHA in labs—Chip remained at the forefront of innovation. He served as Chairman of SEFA during the 1999–2000 rewrite of SEFA 1 and pushed for smarter, more resilient hood designs that could maintain containment even on suboptimal ventilation systems. Along the way, he realized the industry's greatest challenge wasn't technology—it was communication. Fume hoods were being treated as standalone devices, rather than what they truly are: the front-line user interface to a building's ventilation system.

That insight led Chip to found Creative Solutions in 2008, and later Fume Hood Certified, to bridge the gap between design and performance. As a global consultant, he's witnessed firsthand the disconnect between what's installed and what's safe—especially in parts of the world where fume hood education is scarce. Driven by what he saw, Chip became a user advocate, dedicating the last decade to training researchers, laboratory managers, architects, and engineers on how to ensure their fume hoods actually perform.

Now semi-retired but still deeply engaged, Chip continues to champion innovation—especially smart fume hoods that are less dependent on surrounding conditions. He's a sought-after speaker, a standards contributor, and the founder of multiple safety initiatives, all grounded in one mission: to prevent laboratory accidents before they happen.

A. Kaylene Ray

A. Kaylene Ray is a Texas attorney with over four decades of legal experience. Her career has included government regulatory practice in the areas of environmental, workers' compensation, and banking law. In private practice, her work spanned a wide variety of legal areas. This combination of governmental and private practice makes her especially equipped to see liability issues from various perspectives. Kaylene brings an understanding of regulatory compliance, liability, and institutional accountability to the discussion and analysis of laboratory accidents. During the past decade, she has worked with Chip Albright on his mission to make laboratories safer. As co-author of this book, Kaylene examines the intersection of legal obligation and operational failure, helping readers understand how gaps in oversight, training, and enforcement can have consequences for those involved that they may have never considered.

Introduction

Why This Book Matters

Laboratories are inherently dangerous with numerous hazards. Luck only goes so far in accident prevention.

SpaceX uses the term "rapid unscheduled disassembly" (RUD) as a euphemism for an explosion, often used to describe the catastrophic failure of a rocket. Laboratory accidents are also unscheduled events. Everything was ok, until it wasn't.

This book examines the critical topic of laboratory safety by highlighting the ongoing occurrence of preventable accidents. The book's purpose is to explore real incidents, identify what went wrong, the lessons that should have been learned from the incident, and to advocate for systemic changes in laboratory culture and practices.

The Human and Institutional Cost of Laboratory Accidents

We discuss the multifaceted impact of laboratory accidents—loss of life, injury, emotional trauma, institutional liability, damaged reputations, and financial fallout. Accidents are expensive. Prevention is a bargain.

A Pattern of Missed Warnings

We explore how warning signs and near-misses are often ignored, leading to tragic consequences.

The majority of laboratory accidents are preventable, and when they do happen, they should be learning experiences that prompt change and reduce the likelihood of a repeat event.

This book is intended to get you to think. Think about today, tomorrow, next year and a decade from now. A new awareness and food for thought.

Safety is a personal journey, each of us have to blaze our own trail. But hopefully, this book will be a useful guide for you as you work to make laboratories safer and to prevent accidents and mitigate risks.

What happened to Sheri Sangji is a tragedy. Everyone working in laboratories should know her story. While she is not here to tell us in her words, let's learn from what happened and use her story as a motivation to change our practices and prevent another tragedy like this.

This book should be required reading for every university environmental health and safety professional.

We dedicate this book to Sheri.

Section One

The Human Tragedy

The Incident: What Happened on December 29, 2008?

The "Sheri Sangji Story" is a human tragedy of a laboratory accident. A tragedy that most likely could have been prevented had the risks been fully understood and safety procedures been understood and practiced.

A Life Interrupted

Sheharbano "Sheri" Sangji was 23 years old, brilliant, driven, and kind. A recent graduate of Pomona College, she had just begun working as a research associate in a prominent chemistry laboratory at the University of California at Los Angeles (UCLA). Her smile lit up rooms, and her future seemed limitless. But on December 29, 2008, Sheri walked into her laboratory not knowing it would be the last time. Her life would be ending soon.

That afternoon, she was performing a chemical transfer involving tert-butyllithium (t-BuLi), a substance so reactive it ignites spontaneously when exposed to air. She was using a 60 mL plastic syringe—a tool ill-suited for the job—to withdraw the volatile liquid. The plunger slipped free. In an instant, the air-sensitive chemical exploded into flame. Fire consumed Sheri's torso and hands. She wasn't wearing a lab coat. Her synthetic sweater ignited and melted into her skin.

A colleague tried to smother the fire with his own lab coat, but it too caught fire. Flames were eventually extinguished using water from a nearby sink, but the damage was done. Sheri had third-degree burns on nearly half her body. She fought for her life for 18 days before succumbing to the injuries on January 16, 2009.

Her death shook the University and sparked a nationwide reckoning with laboratory safety. How could such an accident happen? But the reality is, when you don't develop a safety culture and practice it, luck will only take you so far.

The conditions in her laboratory told a sobering story. There were no written protocols for handling the hazardous reagent. Sheri had never been formally trained on the material. She had only watched a more experienced researcher once, before attempting the procedure herself. The fume hood she worked in was cluttered, causing her to be too close to the sash opening. The syringe was too small for the volume she was transferring. The needle was too short, forcing her to tilt the container in a risky maneuver. A perfect storm was developing.

Sheri was working alone, unsupervised, on only her second attempt with this dangerous chemical. Though a safety shower was only feet away, in a panic she ran away from it. She had never been shown or practiced what to do in an emergency. After her death, investigations revealed a troubling culture. Safety policies existed but were not enforced. Protective gear was inconsistently used. Hazardous procedures were passed down informally, often without documentation or oversight. And despite inspections noting violations, little changed. No one took responsibility for safety.

Sheri's story became more than an accident report. It became a symbol of the gap between written policy and lived reality in academic research. Her case led to felony charges against her supervising professor—the first of their kind in U.S. academic history.

But more than anything, her story is a human one. Sheri loved books, animals, and her younger sisters. She was thinking of going to law school, possibly to work in chemical safety or environmental justice. She was thoughtful, smart, and full of promise. Everything was ok, until it wasn't. That promise was lost, but it need not be wasted.

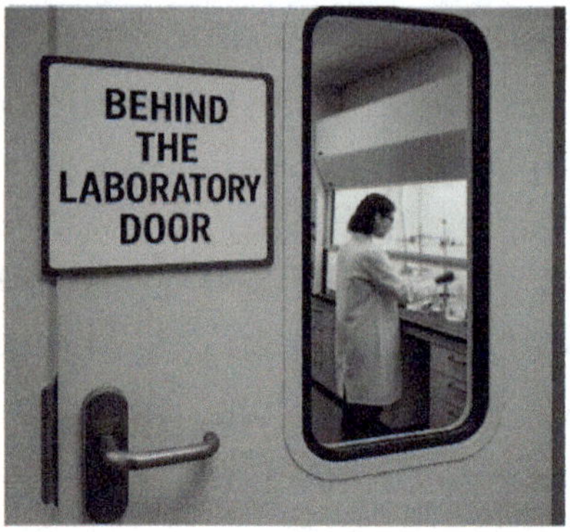

Behind the Laboratory Door - Culture, Supervision, and Safety Norms at UCLA

In the quiet stillness of the UCLA campus during the 2008 winter break, the Molecular Sciences Building should have been a hushed haven for research. But for Sheri, it was another day at the bench—another experiment, another task, another moment in what she hoped would be a meaningful, if temporary, contribution before starting law school. What Sheri didn't know was that she had been placed into a system not built to protect her—a system of academic norms where safety was whispered, not shouted.

Sheri was working as a research assistant to chemistry professor Patrick Harran. From the outside, the Harran laboratory was prestigious. Harran was a rising star in the field of organic chemistry. Inside, it was a high-stakes, high -risk environment driven by ambition, publication pressure, and grant deadlines.

Sheri was expected to execute procedures involving one of the most dangerous reagents in synthetic chemistry—tert-butyllithium—a substance that ignites spontaneously in air. It was only her second time working with it. There were no formal training modules. No written standard operating procedures. No structured onboarding to familiarize her with emergency protocols or the full spectrum of risk. What she didn't know about the hazards killed her.

What she did receive was instruction by another researcher, a post-doctoral researcher who had never read the manufacturer's detailed safety bulletin and did not follow its protocols. He passed along what he knew – what he *thought* he knew. It was science by apprenticeship, a system inherited from generations past, handed down like folklore. Dangerous, incomplete, and unregulated.

UCLA had policies – on paper. Their official Chemical Hygiene Plan delegated responsibility to Principal Investigators, referred to as PIs, like Dr. Patrick Harran, making them the linchpins of laboratory safety. But there was no enforcement, no verification, and no cultural imperative that translated those policies into action. Lab coats, mandated by both University policy and chemical safety standards, were rarely worn. Even after multiple laboratory-related injuries the year before, the University's Environmental Health and Safety (EH&S) Office did not intervene. The gaps were known. They were simply tolerated – part of "the culture." Sheri was wearing a sweater–synthetic, flammable, lethal in the wrong moment. She had gloves and goggles, but no lab coat. No flame-resistant gear. No respirator. She was alone in the laboratory, working near the front edge of a cluttered fume hood, using the wrong size syringe, with the wrong kind of needle. The setup made safe transfer nearly impossible. She had no choice but to work at an awkward angle, tilting bottles and applying uneven pressure. The plunger popped out, and the pyrophoric liquid erupted into flames. But there should have been safety guardrails. She should have been wearing a flame-resistant lab coat. And she should have been aware of the available safety equipment (like the emergency shower) and trained how to properly use it. There should have been an awareness of the risks and a plan for what to do if things went wrong. These simple things would have likely saved her life.

But in the chaos, she didn't head for the emergency shower—just feet away. She had not practiced what to do. There had been no fire drills or emergency responses. No laboratory-wide training in emergency response. A post-doctoral researcher tried to smother the flames with his own lab coat, which also caught fire. Water from the sink was used instead—an improvised act of desperation. Too little, too late. When investigators later asked about the laboratory's safety protocols, they found a paper trail of good intentions and a trail of evidence showing neglect. Dr. Harran admitted there was no formal training provided to Sheri. No assessment of her readiness. No documentation. And when asked why she wasn't given proper PPE, his answer was devastating in its simplicity: "We wear them 90 percent of the time." In reality, lab coats weren't being used, and everyone knew it – including EH&S.

The system failed her – not just the improper syringe or the flames that followed, but the university that allowed this culture to persist. The deeper tragedy is that Sheri's death wasn't a fluke. It was the predictable outcome of an academic system that prized research output over personal safety. A system where legal compliance was treated as a formality rather than a lifeline. Where a 23-year-old was trusted with a pyrophoric chemical, but not with the full truth of its dangers.

Her story must not be treated as a tragic exception. It is a mirror held up to the institutional culture behind the laboratory door – a culture that must change if we are to honor her memory with more than words.

Laboratories Aren't Safer, Don't Be a Statistic

In the aftermath of Sheri's death at UCLA in 2008, it felt like something fundamental had shifted. Her story – a 23-year-old burned to death in a university laboratory – made national headlines. For the first time, an academic researcher faced felony charges for a laboratory safety failure. Institutions issued statements, policies were revised, training modules were updated. And yet, if you look at the statistics (see page 50), very little changed.

So why, after one of the most publicized laboratory accidents in American history, didn't the academic world get significantly safer?

Let's walk through the deeper reasons – through stories, through culture, and through missed opportunities.

Laboratories are not safer for a number of reasons. Changing rules is easier than changing culture. We talk about safety in vague terms that often are not actionable. The focus needs to change to **accident prevention.**

Many commercial institutions have already changed the focus.

However, universities are slow to shift incentives from production to prevention. Grant money and institutional reputation drive the research. Very intelligent people often rationalize that they are smart enough to avoid an accident. "It won't happen to me."

Most young researchers still learn from older ones who weren't trained in safety themselves. Tribal knowledge is lacking and dangerous. Most researchers fail to fully understand the hazards and risks.

Accident "near misses" rarely become institutional lessons.

Sheri's death—though public, painful, and preventable—was still treated by many as a tragic "one-off" rather than a systemic failure.

Safety as an Afterthought

Too often, laboratory safety is treated as a checklist—a compliance task for EH&S departments, not a core part of the scientific process. PPE is an optional inconvenience. Emergency protocols are posted on walls, rarely reviewed. And showers? As some have noted, chemists aren't even trained to use them because they're messy and inconvenient to demonstrate in front of others.

Would you want to fly in a plane with a pilot who had little or no training. Would you give the keys to your vehicle to your child without training? Yet most laboratories are more dangerous than driving. What are we missing here?

When a young woman dies in a laboratory at a prestigious university, the expectation is clear: institutions will wake up. And for a moment, they did. The University of California system revised its safety procedures under legal pressure. Safety committees multiplied. SOPs (Standard Operating Procedures) were reissued. But those were mostly reactive changes.

As Debbie Decker from University of California, Davis put it, it felt like a "tsunami of compliance." Institutions rushed to check the boxes — not necessarily to change the culture. A more formal laboratory education process is needed before people are allowed to work in a laboratory, even with supervision. And as the risks go up, so should the level of training. Forty-plus years of visiting laboratories has shown me failures that take place every day.

For example, at UCLA, even after Sheri's death, a California Division of Occupational Safety and Health (Cal/OSHA)

investigation revealed that other laboratories still didn't enforce personal protective equipment (PPE) use. The culture remained one of "soft rules." Everyone knew what they were supposed to do, but no one made it mandatory. And so, many didn't comply. Something about human nature tends to minimize risk in our minds. We don't get in our car thinking we are going to be involved in a fatal accident. Police or fire personnel don't go to work thinking the unthinkable is going to happen. A private pilot does not get into the plane thinking it will be the last flight. If we can counter this tendency by training and practicing emergency procedures, should something go wrong, we are much better prepared for a positive outcome.

Compliance improved on paper. But day-to-day safety behaviors did not. Safety is no accident; it is an actionable concept that must be taught and practiced every day. Simply checking compliance checkboxes does little to reduce accidents and minimize liability.

The Educational Paradox

It's sobering to realize that 81% of serious laboratory incidents tracked by the U.S. Chemical Safety Board (CSB) in a 2018 report occurred in educational settings. Schools are where young scientists are supposed to learn—not just how to mix reagents, but how to think critically about risk and how to minimize it.

But many universities lack the structure or resources for comprehensive safety education. Students may get a one-hour safety orientation once a year – hardly enough preparation to handle explosive, corrosive, or pyrophoric chemicals.

The reality is that many PIs have made it to their positions with little or no formal safety training. So how can we expect them to pass down what they don't know?

Contrast that with the requirements to work with recombinant DNA or radioactive isotopes. In those fields, researchers must submit protocols, undergo training, and receive committee approval. Why don't chemical hazards receive the same level of scrutiny?

The answer may lie in history. Chemical laboratories have long been governed by an informal code – more tradition than regulation. Changing these practices requires not just new policies, but a transformation in mindset. Most PhDs are very smart people, but their knowledge may be a little myopic especially when they don't understand the importance or risks associated with these tasks. In academia, it's easy to assume that if someone has a degree, they know what they're doing. But degrees aren't training and intelligence isn't a substitute for experience. Sheri wasn't reckless, she was unaware and unprepared.

What happened after the fire ignited is as telling as what led up to it. Witnesses reported that Sheri ran away from the emergency shower instead of toward it. That shower was mere feet away, designed for exactly this type of situation. But Sheri's instinct was to flee, not to douse the flames.

This detail haunts many who have studied the case. It suggests that Sheri had never been drilled on what to do in case of a fire. She hadn't been trained to use the shower. She likely didn't even think of it in the moment of panic.

If she had stopped and used the shower immediately, the outcome may have been different. But you don't do what you haven't been taught – especially in the stress of a life-threatening situation.

Even minor incidents can have severe and tragic outcomes due to lack of training. Security camera footage from a minor run-away experiment shows how such an incident goes from something manageable to a major laboratory fire in just four minutes. The images clearly show the operators had no training as they stood around and discussed what to do as the situation quickly got out of hand.

It is important to understand the risks and hazards of what is happening in the laboratory and to have a plan of what to do if things go wrong. The use of safety equipment must be practiced so that it becomes second nature in an emergency.

It is shocking to find that most people don't know how to use an eyewash correctly.

If universities want to prevent such tragedies, they must dismantle the myths of competency-by-degree, supervision-bytitle, and safety-by-assumption. Real reform begins with institutional accountability, rigorous enforcement, and a cultural shift where safety is not a suggestion, but a shared responsibility.

A Culture of Silence and Complacency

In many laboratories, speaking up about safety can be career-limiting. Graduate students and junior researchers rely on their PIs for recommendations, support, and publications. If the culture minimizes risk or mocks "safety nerds," few are willing to challenge that.

In one story, a researcher recalled trying to advocate for safer procedures and being labeled as "annoying" by peers and supervisors. The message was clear: safety is fine, but not if it slows down the science.

This culture of silence is perhaps the most dangerous element. You can mandate PPE. You can build showers. But, if people are afraid to speak up when protocols are not followed or when "near-misses" occur, unsafe conditions will persist.

Investigators confirmed Sheri was not wearing a lab coat — not even a cotton one, much less a fire-resistant coat as required for handling pyrophores.

This wasn't an isolated lapse. The investigation found that most laboratory members routinely failed to wear PPE, and that Professor Harran was aware of it. He "encouraged" lab coat use, but never enforced it. At the institutional level, UCLA's EH&S department also knew that across campus, PPE use was inconsistent. It had become part of the laboratory culture—a silent acceptance of risk, normalized through daily repetition.

Sheri's synthetic sweater melted into her skin when the flames erupted. A lab coat — especially a fire-resistant one — could have delayed the burn, given her a better chance to escape, and possibly saved her life. But she never had one. There was no record of one being issued to her. How could something this simple happen?

This is not about someone "forgetting" their coat. This is about a systemic breakdown — a workplace where safety measures were known, but not enforced. Where risks were understood by some, but never communicated to others.

The Pressure to Produce—and the Price of Ambition

The academic engine runs on results — publications, grants, discoveries. Graduate students and post-doctoral researchers work long hours, often late into the night and on holidays. Experiments take precedence over safety drills. Principal Investigators (PIs) are not hired or promoted based on their safety record but on their publications and funding.

This dynamic creates an unspoken but powerful incentive to "get the work done," even if it means cutting corners. Sheri's laboratory was trying to scale up a reaction for a high-profile obesity drug project. She had only been on the job for a few months. She was working alone during a university break, working with a reagent she'd barely used once before.

Academia's Apprentice Model: A Broken Chain

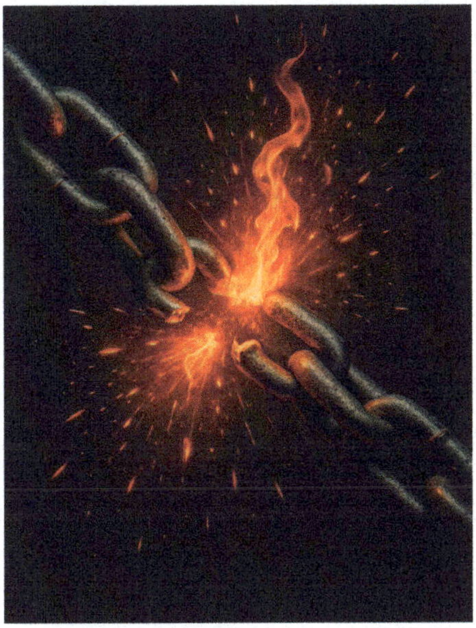

In industry, workers receive formal safety training. They are often required to demonstrate competency before using hazardous materials. In academia, the model is different. Training is informal, passed down from one student to the next — a "tribal knowledge" system.

Sheri was taught how to handle the pyrophoric compound tert-butyllithium by a post-doctoral researcher who later admitted that he himself had never read the manufacturer's safety bulletin outlining how to safely transfer the substance. He didn't follow the bulletin and he didn't teach Sheri to follow it either. Instead, like many in academia, the laboratory relied on what's often referred to as the "apprentice model" - skills passed down informally from one researcher to the next, without written protocols, safety testing or oversight.

This informal "knowledge chain" is dangerous enough in ideal conditions. When applied to a highly volatile, air-sensitive reagent like t-BuLi, it becomes a recipe for disaster. According to Sigma-Aldrich, the manufacturer of the chemical, only fully trained and experienced professionals should handle it. They specifically recommend the use of glass syringes at least twice the size of the volume being transferred and needles that are one to two feet long. Sheri was using a plastic 60 mL syringe with a short 2-inch needle —a clear violation of best practices.

What was the result? An entire laboratory group, under a star PI, using dangerous materials without proper protocols, without real training, and without supervision. Why didn't Sheri know how to properly handle t-BuLi? The answer is as disturbing as it is simple - no one told her.

Failure to Become Institutional Knowledge

A study by University of California, San Diego professor Haim Weizman found that even a decade later, more than 15% of chemistry graduate students had never even heard of Sheri's death. Others vaguely remembered it as "a fire" or "an explosion." Many didn't know the exact reagent or what went wrong. If a tragedy doesn't become institutional memory, how can it serve as a warning?

When a student dies in a university laboratory, it's not just a personal tragedy — it's a public failure. It's a warning, a test, and a moment to act. But year after year, incident after incident, we keep asking the same question: *Why is this still happening?* The real question is: ***Why didn't we learn from the last time?***

Until safety becomes a shared, active, and enforceable priority, from the top administrators to the newest laboratory technician, laboratories will continue to be places where research and danger coexist uneasily.

Conclusion

Sheri Sangji's death was not the result of one bad decision. It was the result of dozens of missed opportunities – by her supervisor, by the laboratory, by the University. Training, supervision, equipment, emergency protocols – every safeguard that could have protected her was either missing or broken.

The cost of laboratory accidents is too often measured in dollars, lawsuits, and damaged reputations. But Sheri's story reminds us that the true cost is always human. It's not just about regulations or protocols – it's about **people.**

Section Two

Property Damage

The Incident: Laboratory Fire at St Andrews University

Not every laboratory accident ends in hospitals or with funerals. Sometimes, the losses are quieter. No one is rushed to the burn unit. There are no memorials, no court cases, no headlines. But that doesn't mean the toll is any less real.

Let's turn now to one of those cases.

It happened in a bright, bustling research facility at the University of St. Andrews in Scotland. The kind of place where science happens every day in shared spaces — benches lined with glassware, humming refrigerators filled with precious samples, notebooks recording months of careful work. Then, one February afternoon in 2019, fire ripped through a third-floor chemistry laboratory. No one was seriously hurt, but the building was left soaked and scarred. Smoke hung in the air. Equipment was ruined. Projects were lost.

It was a different kind of tragedy—one without a casualty but still devastating. While no lives were lost, the human cost remained.

For over a hundred researchers, students, and staff, the laboratory wasn't just a room — it was their second home. Their ideas lived there. Their careers were being built there. When the fire came, it didn't just consume chemicals and ceiling tiles. It swallowed time, effort, hopes, and momentum.

Some had to start over. Others never recovered their work. Some packed up and moved to temporary trailers. Some moved across countries to continue research.

All were affected.

This next story reminds us that even when the flames don't take a life, they can still alter one. Because, science doesn't happen in a vacuum. It happens in human hands. And when something disrupts that process — whether by fire, negligence, or oversight — the scars go beyond the destruction of property. They reach into the hearts of the people who were just trying to do their work.

A Spark That Changed Everything -- A Routine Afternoon Gone Wrong

On February 10, 2019, inside the Biomedical Sciences Building at the University of St. Andrews in Scotland, life was proceeding as usual—experiments humming along, centrifuges spinning, and laboratory members meticulously tracking results in their notebooks and on computers. These were ordinary moments in the life of a research facility. But sometimes, ordinary turns into a catastrophe without warning.

No one saw it coming. One moment the laboratory was just a laboratory, and the next it became ground zero of one of the most disruptive laboratory accidents the University had ever faced. A fire erupted in a third-floor chemistry laboratory and spread to the second and fourth floors. It was later determined that the fire started in a base bath - a container filled with a saturated alkaline solution that is regularly used in synthetic chemistry laboratories for cleaning glassware. What mattered most in that moment was not what caused the fire, but the immediate, unrelenting urgency of the flames.

Thirty firefighters responded. They battled the blaze through the evening, their equipment humming and boots splashing through growing pools of water. Though the actual fire was limited to a specific area, the suppression efforts caused widespread flooding —water soaked the floors, crept into ducts, and ran down stairwells like rivers through a place built for precision, not puddles.

That night, science stopped at St Andrews. But in its wake, something else began.

The First 24 Hours: Smoke, Water, and Uncertainty

In the early evening hours, emergency protocols kicked in. Evacuations were swift. Thankfully, there were no injuries. But as first responders worked to contain the blaze, the damage expanded beyond the visible flames. Smoke infiltrated adjoining laboratories. Water, necessary to smother fire, became its own destructive force.

Within hours, the building was closed. Dozens of researchers were turned away from their life's work. Sensitive instrumentation – some of it worth hundreds of thousands of pounds – sat powerless in rooms turned into swamps. Biological samples that had taken months, even years to cultivate, were suddenly at risk of being lost forever. Fume hoods meant to protect researchers stood idle, their ducts filled with condensation and soot.

University administrators, emergency services, and laboratory directors huddled into the night. Could the building be salvaged? Could research timelines be preserved? Could insurance cover all this? As the sun rose on February 11th, one thing was certain: the fire was out, but the disaster was just beginning.

Reagents, Research, and Ruin

The days that followed revealed the true scale of devastation. While the physical footprint of the fire was relatively small, the secondary effects, especially smoke and water damage were immense.

One of the first challenges was refrigeration. Freezers and

refrigerators, critical for storing reagents and biological samples, had gone without power. Miraculously, salvage crews were able to recover about 90% of contents from 50 to 60 cold storage units. But for some, it was already too late. Heat and humidity had compromised sensitive compounds. Backup systems, though helpful, had their limits.

Long-term experiments were ruined. Years of data were lost. Several laboratories were working on custom-synthesized molecules and unique protein cultures, many of which couldn't be recreated easily – or at all.

Laboratory notebooks, many irreplaceable, were damaged beyond repair. Digital files stored on local computers were unrecoverable without backups. Some laboratories had robust redundancy protocols; others had nothing more than hope.

Hope isn't a strategy. And the loss was tangible.

The Human Cost: Displaced and Disrupted

The fire didn't just destroy equipment and halt experiments – it uprooted lives. Over 125 staff members, from ten different research groups, were suddenly without a laboratory. Graduate students lost thesis experiments. Post doctoral researchers, whose careers depend on timely publications, faced delays they couldn't afford. Some researchers were preparing grant proposals – deadlines that didn't care about fire or flooding.

The University scrambled to respond. Temporary laboratories were constructed using modular buildings supplied by Portakabin and similar firms. These structures, equipped with emergency-grade utilities and safety features, became lifelines. They allowed research to resume, but not without complications. Bench space was limited. Equipment was shared. Collaboration was essential.

Yet, amid the displacement, the academic community rose to the occasion. Peer institutions offered laboratory space. Collaborators abroad opened their doors to St. Andrews researchers. Faculty rewrote project plans and extended timelines.

Amidst it all, the hum of recovery replaced the silence of the shutdown.

Psychological Toll: The Fire You Can't See

Disaster changes people. For many at St Andrews, the fire wasn't just about work lost — it was about emotional resilience tested. The sudden void left by inaccessible laboratories felt deeply personal. Some researchers found themselves grieving — not just their data but their direction. There were no casualties, but that doesn't mean there were no injuries. Anxiety, stress, and uncertainty set in. Would students finish their degrees on time? Would grants be pulled? Would jobs be lost?

The University offered counseling and mental health support, acknowledging that safety also includes psychological well-being. That move, small as it may seem, reminded the entire community that people matter more than projects.

Financial Fallout: More Than Just Bricks and Beakers

Estimates for direct repairs ran into the millions. Rebuilding laboratories, replacing damaged systems, and ensuring new infrastructure met upgraded safety standards was no small feat. Insurance covered some of it, but not all.

Indirect costs mounted quickly: lost productivity, delayed publications, extended salaries for staff unable to perform their core duties. Some grants had to be renegotiated. Others were lost altogether. Collaborations stalled. The chain reaction, much like a chemical process, was swift and expansive.

The incident sparked conversations across the United Kingdom higher education sector: were laboratories really financially prepared for a worst-case scenario?

Rebuilding: From Ashes to Innovation

By late 2019, the first modular laboratories came online. They weren't perfect, but they were functional. Work resumed, cautiously at first, then more confidently. Researchers adapted. Students graduated. Papers were eventually published.

Meanwhile, reconstruction of the Biomedical Sciences Building pressed on. Engineers and safety consultants seized the opportunity to rebuild not just stronger, but smarter. The new building would include new fire suppression systems, improved chemical storage, upgraded fume hoods and airflow monitoring, better emergency lighting and improved signage.

When the refurbished and modernized laboratories officially reopened in 2023, they bore little resemblance to their former selves. These were not the same rooms where the fire started. These were spaces born from lessons — hard-won, painful, necessary lessons.

Preparing for What Shouldn't Happen

If the fire at St Andrews taught the scientific world anything, it's that **prevention and preparedness are not luxuries — they are necessities.**

In every department meeting post-fire, disaster preparedness became a major topic: inventories, backups, and risk audits became high priorities. Laboratories began treating digital and physical backups as sacred.

The solidarity shown — both within the University and from external collaborators — proved that no laboratory is an island. The network of support carried the community through.

Graduate students and post-doctoral researchers bore the brunt of the disruption. Future planning now includes better safeguards for early-career researchers, from alternative laboratory access plans to mental health provisions. Redundant systems were implemented. From cold storage redundancy to cloud-based data backups, the fire triggered investments in resilient infrastructure.

Reflections: What a Fire Can Teach

The St Andrews fire was, in every way, a catastrophe. It erased years of work in minutes and challenged an entire institution. But it also laid bare truths that often lie hidden: that laboratories are fragile, that safety is more than a checklist, and that culture matters just as much as chemistry.

When the ribbon was cut on the newly restored building, it wasn't just a celebration of drywall and ductwork. It was a moment of collective triumph. The fire had tested them — and they had passed, not by avoiding damage, but by rising from it.

The scientists of St Andrews returned to work not as they were before, but as people who knew that even in the most precise environments, unpredictability reigns. Their story isn't just about a fire. It's about what we do after the flames are out.

What was the financial cost?

The project to restore the Biomedical Sciences Building at St. Andrews cost an estimated £21 million. While the UCLA incident took Sheri's life, there was little property damage. At St Andrews, there was no loss of life or injury but massive property damage and research loss. Both were tragedies. Safety is a vague overused word. The focus should be on "**accident prevention**". While it is impossible to prevent all accidents, we want systems and processes that mitigate their severity. No one wants to be a statistic. As we continue this journey to understand the cost of these tragic accidents, we should consider the cost of preventing them. Wouldn't it make sense to spend a fraction of the cost of an accident on making laboratories safer and protecting the health and wellbeing of laboratory users? The old adage, an ounce of prevention is worth a pound of cure couldn't be truer.

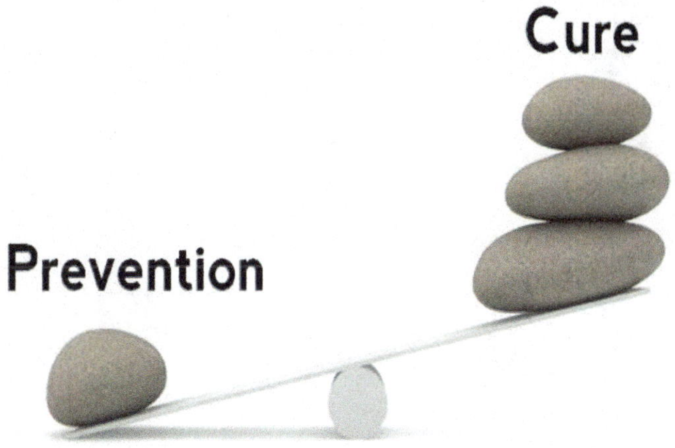

Have A Plan,
Don't Count on Luck
Hope is Not a Strategy

Section Three

Who is responsible?

Legal Liabilities and Issues Unresolved

Accidents happen, some small, some noteworthy. But when accidents happen, who is responsible/liable? This is a topic often overlooked or not talked about.

Stakeholders

The number of stakeholders related to laboratory accidents is broader than you might realize. The following chart focuses on fume hoods, but as primary safety device in laboratories, the stakeholders are very similar to general safety. You can see from this chart, that the list of stakeholders is large and wide-reaching. And when things go wrong, the finger of blame can be pointed in all directions.

You don't have to be solely responsible to be held liable. Joint and several liability can pull you into a legal battle even if others are also found at fault.

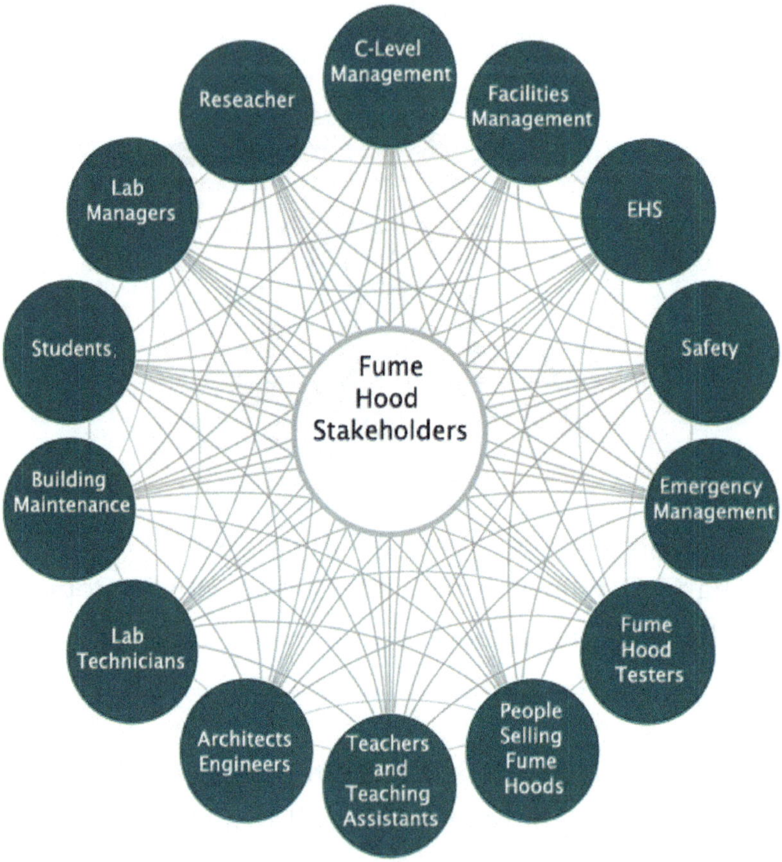

There are three basic elements in the establishment and use of a laboratory, with different stakeholders involved in each. They work together like a three-legged stool, unless you address all three elements, accidents are more likely to happen.

Element 1 - Design of the space, this involves Architects and Engineers.

Element 2 - Operation of the space, this involves owners, administrators, facilities management, EH&S, maintenance.

Element 3 - Use of the space, this involves laboratory managers and supervisors and the people using the space.

Unless equal attention is paid to all three elements, the likelihood of an accident happening is increased.

Safety/Accident Prevention is a dynamic system and the number of variables that occur in a laboratory setting make it particularly complex. Let's compare it to something a bit more familiar - driving.

We are going to drive between New York City and Los Angeles. We will be taking all types of roads - toll ways, expressways, interstate highways, state highways, and even rural roads. We will encounter different driving conditions, different speed limits, and situations like construction and school zones. In those 2,800 miles and 50 hours of driving, we encounter many situations with little standardization. As we progress from state to state we will see many different conditions. Laboratories are much like that. What they do varies dramatically. Safe driving involves putting what we encounter into context. We may be able to safely drive 80 MPH on a remote rural interstate, but we need to slow down in town for the school zone. Each state and city has its regulations for what they feel is safe. We could write an entire book on what we encountered on this road trip -- both good and bad. And remember, we are not on this journey alone. There are many others we must share the roadway with. Many types of vehicles, many levels of driving experience, and those that are distracted or that believe that the regulations don't apply to them.

Laboratory safety is similarly complex. It is not a one-size-fits-all scenario. And it is not only the infrastructure and the equipment that affects safety; it is also how the people interact with the space and others around them. One distracted person can put the whole laboratory at risk.

It is often easier to look the other way than to tackle this head-on. Following a SOP might make us compliant, but "soft rules" and lack of enforcement put us at risk. There is a huge difference between **safety compliance** and **accident prevention**. A fully compliant laboratory can still have an accident, but compliance with safety rules does have an impact on liability. Safety is a top-down responsibility, but to really be effective it must be a bottom-up process. There must be buy-in at all levels.

If you are reading this book, chances are you could be held accountable and liable in case of a serious accident. So, how can we mitigate risks? It all starts with education. Understanding the hazards and the risks and determining how significant they are is key. Again, often context helps. A house painter of a one-story house has a certain level of risk, but if the painter is painting a 25-story building, the risks are entirely different. To keep both equally safe, requires entirely different types of systems. Laboratories are not much different. Even on the same campus or in the same building, each laboratory will have different risks. In a teaching laboratory, the risks can be different based on whether the demonstration is being done by a PI or a teaching assistant.

The Legal Background

Now let's put the legal hat on. Understanding the bases for liability dangers is helpful. Generally, these include: regulatory violations, civil litigation, workers' compensation claims, and criminal prosecutions. Here are the details.

Regulatory Violations

Determining what regulations apply to university laboratories is a complicated analysis. The federal regulations of the Occupational Safety and Health Administration (OSHA) contain a number of specific standards that apply to laboratories. These include The Occupational Exposure to Hazardous Chemicals in Laboratories standard (29 DFR 1910.1450), commonly referred to as the Laboratory Standard. This provision requires designation of a Chemical Hygiene officer and adoption of a written Chemical Hygiene Plan (CHP). Other sections require the provision of personal protective equipment (PPE), eye and face protection, respirators and hand protection. The Hazardous Communication Standard (HCS) (29 CFR 1910.1200) requires evaluating the potential hazards of chemicals and communicating information concerning those hazards and appropriate protective measures. This requirement is based on the concept that persons working in laboratories have both a need and a right to know the hazards and identities of the chemicals they are exposed to. All of these and various other OSHA regulations have as their goal to protect and assure the safety and health of laboratory workers.

Federal OSHA standards only apply to relationships established between employers and employees and therefore do not extend to students. Additionally, the Occupational Safety and Health Act (OSH Act) specifically covers private employers. In the university laboratory context this is important because many colleges and universities are public entities. Does this mean that laboratories in these public colleges and universities are not required to comply with OSHA standards? Well, as lawyers often respond, "it depends." It depends in what state they reside.

States are allowed to develop and operate their own OSHA-approved State Plan. Such a plan must offer no less protection than is afforded by the federal OSH Act. There are currently 22 states or territories that have such State Plans that cover both private-sector and state and local government workers. There are seven State Plans that cover only state and local government workers. Therefore, *public* colleges and universities may be subject to State Plans, but all *private* colleges and universities are subject to either an OSHA-approved State Plan or the federal OSH Act.

For those institutions covered by either the OSH Act or a State Plan, violations can be costly. As of this writing, a serious violation (defined as one that can cause illness or accident resulting in severe injury or death) carries a penalty of $1,221-$16,550 per violation and that penalty can be doubled for each day the violation is not abated. Willful or repeated violations carry a penalty of up to $165,514 per violation.

In the 2008 UCLA case the university was fined $31,875 by the California Division of Occupational Safety and Health (Cal/OSHA) on four violations of the law.

Tort Litigation

Lawsuits are a huge drain of time and resources and a big distraction from the things that need your focus. Generally, when a civil action is filed in response to a laboratory accident, it will be a lawsuit alleging that a tort has occurred.

A tort is an act or omission that causes harm to persons or property. Four elements must exist to prove tort liability:

1. a legal duty of one person to another;
2. a breach of that duty;
3. an injury - physical or monetary; and
4. the breach of duty is determined to be the proximate cause of the injury.

Higher education institutions have a responsibility to protect students from harm. In the case of laboratory incidents at colleges and universities, the duty involved is typically the duty of an instructor to supervise and train. Courts have generally established the standard of care as, what a reasonable person would do to prevent foreseeable injuries. Unfortunately, there are very few court cases involving injuries in college and university laboratories to refer to in analyzing the facts that courts find relevant in establishing the extent of an instructor's duty to students in the laboratory. It is clear however, from reading cases regarding grade school and high school laboratory incidents, that the duty of care will vary based on the age and experience of the persons involved. So, it follows that the duty to supervise a high school chemistry student, verses a college freshman in Chemistry I, verses a PhD candidate, will be judged differently. In addition, closer supervision and more extensive instruction would be expected for experiments and research using materials or equipment that are more dangerous. If an injury does occur in a laboratory, an instructor also has a duty to provide first-aid.

Likewise, a court may consider age and experience in determining whether a breach of the duty owed has occurred. Leaving a PhD candidate alone to perform routine tasks may be reasonable while the same act may be a breach of duty when an unexperienced freshman student is involved.

Proving that a breach of duty was the proximate cause of an injury is necessary to establish liability. While there is little guidance from the courts on this relating to laboratory situations, factors such as unexpected or spontaneous events which prevent supervision from intervening, or the actions of another student, could complicate the connection of the instructor's failures to the injury.

Common defenses raised by defendants in tort cases include arguments that the injured person was also at fault. This is called contributory negligence and is shown when the person injured fails to exercise ordinary care for his own safety and this failure is part of the proximate cause of the injury. For instance, when a student fails to follow well-known instructions or engages in reckless actions. Similarly, where an experienced, well-trained, graduated student chooses to ignore established protocols and is injured as a result of those choices, a defense of "assumption of the risk" could be argued.

As you can see, all of the elements necessary to prove tort liability are very fact specific and the propensity of these cases to settle out of court gives us little guidance on how courts and juries will analyze and decide these issues. In addition, settlements often include non-disclosure agreements (NDAs) which prevent the details of the incident and the settlement terms from becoming public.

Public colleges and universities, when sued in a tort claim, will inevitably argue that they are immune from suit as a state institution. This defense of sovereign immunity is based on the various doctrines and statutes that provide federal, state or local government immunity from tort-based claims. The theory of sovereign immunity dates back to English common law and has evolved over time. It is now largely a statutory concept with each state adopting its own version of a tort claims act. Generally, these acts include an absolute immunity waiver, a limited waiver for certain specific types of claims, or a general waiver with certain defined exceptions. A review of the law in the state where an institution resides and where an accident occurs is necessary to determine how it affects related litigation. In the context of litigation against colleges and universities and their employees, the issue of sovereign immunity from tort claims is only relevant to public institutions.

Workers' Compensation

The U.S. workers' compensation system began at the turn of the

20th century when Wisconsin passed the first comprehensive workers' compensation law. Other states followed with the last to pass workers' compensation legislation in 1948. The basic premise of these laws was that, in exchange for a guaranteed payment if injured on the job, workers gave up their right to sue their employer over workplace injuries. Each state's workers' compensation law is different and has its own definition of who qualifies for coverage. Generally, compensation to injured workers is provided in the form of wage-replacement for the period of disability, a lump-sum payment in the case of any permanent disability, and payment of medical and rehabilitation costs. The system is employer-funded by the purchase of commercial insurance or setting up a self-insurance account.

Workers' compensation is a benefit only available to employees. In the academic laboratory world, one of the most confusing and contested issues is whether an injured laboratory student is an employee of the college or university for workers' compensation purposes. (This determination is also important in assessing OSH Act applicability.) This determination can be a complicated analysis given the various roles students take on in the academic laboratory setting. Employment status is a fact and law-specific determination - getting a pay check does not necessarily resolve the issue. Factors such as who was paying the student for the work being done at the time of the incident come into play. Graduate students in particular are often given stipend payments, fellowships and grants that may not qualify them as employees for workers' compensation purposes. Some cases have arisen on the issue of what a graduate student was working on when an injury occurred with universities contending that

PhD research was for the student's benefit, not that of the university, disqualifying the student for workers' compensation benefits.

Even with the issues that arise, workers' compensation provides benefits that shouldn't be discounted. The employer gets relief from tort liability, the employee gets relatively quick, equitable and predictable, no-fault compensation, and the system includes rehabilitation incentives. Tort litigation is lengthy, expensive and unpredictable for all parties.

Criminal Charges

While tort law aims to compensate individuals who suffer harm, criminal law aims to punish individuals who violate the law. You wouldn't expect to face criminal actions in a laboratory incident, but that's exactly what happened in the lethal 2008 UCLA laboratory incident. Three years after the incident that killed laboratory research assistant Shari Sangji, the Los Angeles District Attorney charged the chemistry professor in charge of the research and the regents of the University of California with four felony counts of willful violations of California's labor code. Conviction on all counts carried a penalty of up to four and a half years in prison. This was the first criminal prosecution in the U.S. for an academic laboratory accident and it shocked the scientific community.

The board of regents soon settled with the District Attorney agreeing to accept responsibility for the conditions under which the laboratory was operated, to create a scholarship in Sanji's name, and to establish a program of safety training and compliance in all their laboratories. In exchange, the District Attorney dismissed the charges. The professor in charge however, did not settle and continued a legal battle with the prosecution for years, until he reached a settlement with the District Attorney in May of 2014. The details of the settlement included 5 years of probation, teaching obligations, community service and a $10,000 donation to the burn center where Sangji died, after which the charges would be dismissed. Like so many of the tort cases, the settlements left unanswered the question of whether a laboratory chief and a university can be held criminally responsible for a death in their laboratory.

Criminal charges bring the consequences of a laboratory accident to another level. Being fined or reprimanded or even found liable in a civil court is one thing, but the possibility of incarceration is something quite different.

Unresolved Questions

The common scenario in laboratory accident incidents is that civil suits are settled, sometimes with NDAs, plea agreements are entered into in criminal actions without a trial, workers' compensation insurance pays for medical costs in a no-fault environment or institutions find ways to pay medical costs outside of workers' compensation. This reality leaves an absence of documentation that makes it difficult to thoroughly research these incidents. In addition, the weak reporting requirements for laboratory incidents make available statistics less than reliable.

Many questions related to responsibility in laboratory accidents are left unanswered. What is the standard of duty? Who has this duty? When is the duty breached? Who is covered by workers' compensation insurance? When is an institution or a person in charge of a laboratory guilty of a crime?

Amid a sea of uncertainty one thing is clear, if the accident had been prevented or mitigated in the first place, there would be no reason to ask these questions.

Accident prevention is the answer.

Section Four

The True Costs of a Serious Laboratory Accident

Financial, Institutional, and Human Impacts

When a laboratory accident occurs, the immediate instinct is often to assess the physical damage: broken glassware, scorched benches, destroyed samples. But these visible losses are merely the tip of the iceberg. The true costs of a serious laboratory accident are far-reaching and devastating — affecting people, research, reputations, and the very mission of science itself.

The Human Toll: Irreplaceable Loss

The most devastating cost of a laboratory accident is, of course, the human one. In the case of Sheri Sangji, in addition to her terrible pain and suffering, her death shattered lives and left scars on colleagues, friends, and fellow researchers who still carry the memory of that day. For the Sangji family, justice meant ensuring no other parent would endure what they had. But grief never closes the ledger — it remains an ongoing emotional debt.

Legal and Financial Fallout

After Sheri's death, UCLA and Professor Patrick Harran faced unprecedented criminal charges — the first ever filed for a laboratory accident in an academic setting. Cal/OSHA fined UCLA $31,875 for four violations. But that was just the beginning.

Legal fees for the University exceeded $4.5 million, according to public records, a sum paid with public funds.

In addition to regulatory fines, legal defense, and settlements, the university faced enormous costs in reputation management and compliance enforcement. Deferred prosecution agreements, public investigations, and years of legal scrutiny forced UCLA into reactionary safety overhauls.

Lost Productivity and Research Setbacks

When a laboratory accident happens, research halts — sometimes for days, sometimes forever. Sensitive equipment is damaged, experiments are invalidated, and years of data may be lost. At St. Andrews University, the direct cost of the 2019 fire included millions in physical repairs — but the indirect cost was staggering: over 125 researchers were displaced, experiments destroyed, and timelines upended.

Projects relying on unique reagents or biological samples, sometimes cultured for years, were lost. For doctoral students and early-career scientists, this can mean more than just an inconvenience — it can mean derailment of an entire academic path.

Reputational Damage

For a top-tier institution like UCLA, being associated with fatal negligence drew intense media scrutiny. What should have been a bastion of cutting-edge research became the face of a preventable tragedy. The public backlash, academic outcry, and internal damage control severely tarnished the University's brand. Worse, it cast doubt on the integrity of academic laboratories as a whole, sparking national conversations about safety oversight.

Damage to reputation doesn't just affect donor support or rankings. It can compromise grant applications, faculty recruitment, and public trust in science.

Psychological Cost to Staff and Students

The trauma endured by laboratory mates, supervisors, and emergency responders is difficult to quantify but very real. At UCLA, post-doctoral researchers like Dr. Wei Feng Chen tried to extinguish the flames engulfing Sheri with their own lab coats. Witnessing such horror, especially in a professional setting where safety should be paramount, leaves psychological wounds.

Institutions rarely provide adequate mental health support after such incidents. The aftermath is filled with guilt, fear, and an erosion of confidence. Students question their own training and decisions. Some leave research altogether.

Cost of Emergency Response and Recovery

Fires like the one at St. Andrews require massive coordinated emergency response efforts. Over 30 firefighters fought the flames through the night, and water damage from suppression efforts extended well beyond the original blaze. Laboratories were flooded. Sensitive areas like cell culture rooms were contaminated. Structural reinforcements and asbestos checks followed, leading to months of downtime and disruption.

Even with insurance coverage, the recovery effort involved extensive logistical coordination and investment in temporary laboratories to prevent long-term loss of productivity.

Cost of Prevention vs. Cost of Inaction

One of the most striking contrasts comes in comparing what *could have been spent* on prevention. A flame-resistant lab coat costs under $100. Proper training takes hours, not months. A simple SOP outlining safe handling of pyrophoric reagents might cost a department a few days of documentation time.

In Sheri Sanji's case, had her laboratory enforced standard PPE usage, properly trained her on the handling of tert-butyllithium, and ensured appropriate oversight, the accident would likely never have occurred.

As noted in the *Lab Manager* article "The Cost of Lab Safety," proactive investment in safety saves vastly more in potential losses – from lawsuits and downtime due to destroyed research and shattered lives.

Section Five

Lessons We Should Have Learned, But Didn't

If the fire at St Andrews taught us how fragile our systems are, and the tragedy at UCLA reminded us how unforgiving the cost of human error can be, then we now find ourselves at a critical junction—a place where reflection must turn into resolve.

Despite headlines, lawsuits, upgrades, and investigations, the laboratory world continues to operate with an unsettling déjà vu. Accidents echo one another. Warnings fade. Best practices remain optional instead of foundational.

The Statistics

The Chemical Safety Board (CSB) tracks and reports on the most serious accidents that involve substantial property damage or serious injury or death. The purpose of this is similar to that of the National Transportation Safety Board (NTSB), to uncover the root cause of the accident and propose corrective actions to minimize the likelihood of it happening again.

Sheri's death was a wakeup call. Most in the scientific community heard about it. UCLA did a lot to address the atmosphere that led up to the tragedy. Millions of dollars were spent on corrective actions. As time passed, people asked a difficult question. Are laboratories safer since this very publicized tragedy? To examine this issue, the CSB published a special report (www.fumehoodcertified.com/CSB) and it looked at accidents before and after the UCLA incident.

CSB

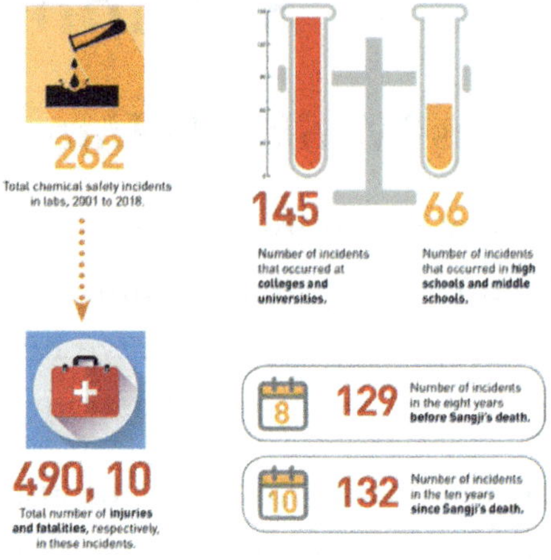

How many is too many?

81% of the accidents were in an educational setting

The CSB statistics showed that between 2001 and 2018 there were 262 reported laboratory incidents, 490 injuries, and 10 deaths. In the 8 years before Sheri's death in 2009, there were 129 serious accidents. In the 10 years after, there were 132 serious accidents. So, the number of accidents didn't dramatically decline. From this data, it appears that the safety and accident prevention message is not effective in reducing the number of laboratory incidents. If you follow the link above and download the total report, you'll see that the accident list includes some of the most prestigious research universities. This is concerning. Yet, the most shocking fact is that 81% of the accidents occurred in an educational setting. Why?

Academic vs. Commercial Laboratories

While statistics can be complex and debated due to reporting methodologies, they show that colleges and universities tend to have more laboratory accidents than commercial laboratories.

We can speculate on the reasons for that.

A 2022 study of safety inspections in research laboratories at a large university looked at the role of PIs in non-compliant practices. This study found that PIs who were well-funded, high status, tenured researchers disengage from the regulatory system often using reactions to regulatory feedback the study describes as, deny, dispute and delegate. More specifically, these PIs tend to deny the relevance of safety regulations and requirements, dispute the fact that the situation is a real safety problem and then cope with the situation by delegating responsibilities to subordinate laboratory personnel. This model of behavior was exemplified by Dr. Harran's actions in the 2008 UCLA incident. Such attitudes in academic research laboratories are passed on to others and become systemic norms. The decentralized and hierarchical structure of academic institutions make it challenging to establish clear lines of safety accountability throughout the laboratory. This is less prevalent in commercial laboratories where the lines of responsibility for safety are more clearly defined and enforced by dedicated safety officers.

Another study conducted in 2023 using responses from university students in a Chinese University concluded that these students were not sufficiently aware of laboratory safety. Recommendations included establishment of laboratory behavior norms, creation of extensive and mandatory laboratory safety education and training courses and establishment of a laboratory accident database.

Training in academic settings is not always as comprehensive or tailored to specific hazards as in industry where rigorous, ongoing safety training is common. Academic researchers are less likely to conduct formal risk assessments before experiments, whereas such assessments in commercial laboratories are standard and mandatory, often requiring formal written documentation and approval.

While commercial laboratories typically employ experienced professionals who are extensively trained in the specific protocols and hazards of their research, university laboratories have a constantly rotating population of students and researchers who have varying levels of experience and training. As students learn and experiment often for the first time, the risk of human error due to lack of attention to detail and over confidence increases. Sometimes the attitude of academic laboratory personnel is that the students, "will only be here for a few hours, so why waste time on safety training."

The type of work conducted in laboratories can also affect safety. While academic laboratories typically work with smaller quantities of hazardous materials, the diversity and exploratory nature of the research can produce a wider range of hazards. Industry may handle larger volumes of hazardous materials but usually work with a more limited focus using highly structured and proven processes.

In addition, the disparity in funding between academia and private industry can't be discounted. University research using funding sources such as grants, operate on more constrained and uncertain budgets which can affect the availability of the latest safety equipment or even standard PPE.

These are some of the differences in the academic and commercial worlds which can explain differences in the numbers of laboratory accidents reported. Recognizing these distinctions provides yet another lesson - observe what is effective in making laboratories safer and work to emulate those practices. Commercial laboratories can provide examples of good safety practices.

It seems that the core values that drive commercial labs differ from those in education. Maybe it is time for a paradigm shift.

The Essential Role of the EH&S Department

We expect that many of you reading this book are from EH&S departments and that many of you are from universities or other educational environments. Given that, we want to take a brief look at the responsibilities of EH&S.

The Environmental Health and Safety (EH&S) department at a university plays a critical role in protecting the health and safety of students, faculty, staff, and visitors, as well as ensuring regulatory compliance. Their responsibilities cover a wide range of functions across academic, research, and administrative environments.

Here's a breakdown of core EH&S roles at a university:

□ 1. Laboratory Safety

Train researchers and lab staff on chemical handling, personal protective equipment (PPE), and emergency procedures.

Inspect laboratories for compliance with safety standards (e.g., chemical storage, fume hood performance).

Oversee hazardous materials use (e.g., chemicals, biological agents, radioactive substances).

Ensure adherence to OSHA's Laboratory Standard (29 CFR 1910.1450) and other applicable codes.

🔥 2. Fire and Life Safety

Conduct fire drills, evacuation planning, and fire extinguisher training.

Inspect fire alarms, sprinkler systems, and emergency exits.

Ensure compliance with NFPA codes and local fire authority regulations.

♻️□ 3. Environmental Protection

Manage the collection, storage, and disposal of hazardous waste.

Ensure compliance with EPA, state, and local regulations.

Oversee programs for air and water quality, stormwater runoff, and spill response.

💼 4. Occupational Health and Safety

- Perform risk assessments for workplace hazards (e.g., noise, ergonomics, indoor air quality).
- Investigate injuries and incidents, and implement corrective actions.
- Coordinate with HR and Workers' Comp programs on employee health concerns.

☣□ 5. Radiation and Biosafety

- Maintain compliance with radioactive material licenses and radiation safety programs.
- Oversee Institutional Biosafety Committees (IBCs) for biological research compliance (e.g., NIH Guidelines).
- Monitor and manage biological safety cabinets, autoclaves, and sharps disposal.

📚 6. Training and Education

- Provide mandatory safety training (e.g., laboratory safety, bloodborne pathogens, hazard communication).
- Maintain training records and refresher schedules.
- Promote a safety culture through outreach, safety campaigns, and student engagement.

💼 7. Emergency Preparedness and Response

- Develop and maintain emergency response plans.
- Coordinate with campus security and local responders.
- Serve on incident response teams during fires, spills, natural disasters, or pandemics.

⚖️□ 8. Regulatory Compliance and Audits

- Ensure the university complies with OSHA, EPA, DOT, CDC, NIH, and local regulations.
- Prepare for internal and external audits (e.g., EPA inspections, accreditation bodies).
- Maintain up-to-date documentation and reporting systems (e.g., SDS, incident logs).

□ 9. Risk Management Support

- Support insurance and liability assessments.
- Assist with design reviews for new labs or buildings from a safety perspective.

- Identify and mitigate potential risks before accidents occur.

In short, EH&S functions as a compliance officer, educator, emergency planner, and safety advocate all rolled into one. That role is essential in keeping the university community safe, compliant, and prepared – especially in high-risk research settings.

While the list may seem daunting, it is reality. And if you work in EH&S you will receive a lot of scrutiny in the event of a serious accident.

Knowledge is Power

The more you know and understand about the hazards and risks of laboratory work, the better prepared you will be to improve laboratory safety. EH&S is somewhat the missing link in a broken chain. Educating and training those in the laboratory, many of whom have had no formal training, allows them to participate in their own safety journey.

Lessons

This next section examines the core lessons about training, PPE, hazard communication, risk ownership, leadership, and culture. Seeing the consequences of accidents at other laboratories should have taught these lessons, and yet, many remain neglected. We must ask not only *what went wrong*, but *why didn't we fix it?* The answers are uncomfortable, but necessary to fully understand what actions need to be taken.

Due to the type of work being performed, laboratories are dangerous places. Accidents will happen, so we must plan on that. Making people aware of the hazards is the first step in preventing accidents. Having a plan on what to do when something goes wrong is the next step in minimizing a negative outcome. Then practicing what to do helps keep it top of mind.

A common response to the need to invest in safety and accident planning is that "we don't have the budget for that." But, do you have the budget for an accident? Safety and accident prevention shouldn't be luck; it should be a planned and practiced event. Look at the damage these two accidents caused. And, they are just highly publicized examples. Accidents happen every day. While most are minor and the costs are minimal, each has the potential to be serious and maybe even deadly. Major accidents can cost in the millions and typically insurance only covers a portion of the cost.

The UCLA laboratory fire that killed Sheri Sangji and the University of St. Andrews fire that destroyed research infrastructure were both tragic events – but they were also preventable. The lessons we should have learned from these two incidents are multifaceted. Here are the core takeaways.

Training Is Not Optional – It's Foundational

- **UCLA:** Sheri Sangji received no documented training on how to handle tert-butyllithium. The laboratory relied on informal "apprenticeship-style" instruction without ensuring comprehension or retention. There were no formal SOPs or verifiable safety protocols specific to the hazardous reagent.
- **Lesson:** Training must be structured, documented, and competency-based – especially for procedures involving high-risk chemicals or equipment. Laboratories should never assume prior knowledge.

PPE Compliance Must Be Enforced, Not Merely Encouraged

- **UCLA:** Sheri wasn't wearing a lab coat, flame-resistant clothing, or gloves rated for chemical fires. The laboratory culture did not enforce PPE use, and EH&S knew that many researchers across campus routinely ignored these rules.
- **St. Andrews:** While no injuries occurred, the fire raised concerns about equipment placement, safety infrastructure, and preparedness.
- **Lesson:** Safety policies must be enforced with zero-tolerance for non-compliance. PPE should be considered essential equipment, not a personal choice.

Hazardous Materials Demand Specialized Handling

- **UCLA:** The improper use of a large syringe, failure to follow manufacturer safety bulletins, and using the wrong equipment (short needle instead of long cannula) led to a deadly fire.
- **Lesson:** Hazardous materials require precise protocols, specialized equipment, and supervision by experienced personnel. Manufacturer guidance must be treated as minimum safety standards.

Safety Culture Must Be Lived, Not Just Filed

- **UCLA:** There were documented violations: no Injury and Illness Prevention Program, no laboratory-specific SOPs, and no laboratory-wide safety audits. The safety culture was reactive, not proactive.

- **St. Andrews:** Although basic protocols were in place, the disaster emphasized the need to prepare for *when*, not *if*, an accident happens.

- **Lesson:** A culture of safety requires leadership buy-in, ongoing dialogue, accountability at all levels, and regular drills that treat safety as part of daily operations. Safety is not a checklist - it's a mindset.

Preparedness Reduces Damage and Downtime

- **St. Andrews:** Damage was largely from smoke and water used during firefighting. A strong disaster recovery effort (e.g., backup of data, salvage of samples, modular laboratory deployment) helped reduce research loss

- **Lesson:** Laboratories need business continuity plans that include inventory audits, digital backups, alternative laboratory space agreements, and clear emergency protocols.

Human Costs Go Beyond Physical Injury

- **UCLA:** Sheri's death was a tragedy that still resonates as a symbol of institutional failure and disregard for young researchers' safety.
- **St. Andrews:** Though no one was hurt, over 100 researchers were displaced, and long-term research projects were delayed or lost.
- **Lesson:** Safety incidents affect not just infrastructure, but morale, careers, and the mental health of entire research communities.

The Cost of Safety Is Less Than the Cost of a Single Incident

- As noted in "The Cost of Lab Safety" article: Investing in proper safety practices is far cheaper than the combined costs of legal liability, reputational damage, lost data, property damage and injury or death.
- **Lesson:** Budget constraints cannot justify cutting corners on safety. It's a strategic investment in sustainability and risk management.

We Keep Repeating the Same Mistakes

- **UCLA:** Despite the high-profile nature of the UCLA case, a 10-year follow-up study revealed that many students didn't even know about the incident, and many laboratories continue unsafe practices. The same systemic issues – lack of training, informal practices, weak enforcement – keep showing up.

- **Lesson:** Documentation and reporting of laboratory incidents and "near-misses" is important. Passing this information to laboratory supervisors and users allows them to learn from past experiences and avoid future mishaps.

Conclusion: A Moral and Financial Imperative

Serious laboratory accidents don't just burn flesh or buildings – they incinerate futures, reputations, and institutional integrity. The costs are high, and every accident is a story of opportunities missed. Safety is a complex concept that should be approached with a highly structured process. If we shift the focus from "Safety" and "Compliance" to "Accident Prevention" the task will become more manageable.

Preventing the next tragedy doesn't require new laws or technologies. It requires the will to prioritize safety as a value, not just a compliance checkbox. Because the true cost of a laboratory accident is never just what you see on a balance sheet – it's also measured in lives altered, careers destroyed, and the enduring echo of a scream that should have never been.

Epilogue: The Fire You Can't See

The beeping of machines has stopped. The hallway is quiet now. Families have gone home, some with grief too deep for words, others with questions that will never be answered. The headlines have faded, the lawsuits are archived, and the burned walls have been rebuilt. But this story – this pain, this lesson – is not over. It waits, just beneath the surface of our routines. Waiting, because we didn't learn. Not really.

Sheri Sangji walked into UCLA's laboratory with no idea it would be the last real step in her life. She did not know that the task she was assigned—one she was not fully trained for – was a chemical time bomb. She did not know the syringe would fail, or that fire would consume her synthetic sweater like dry kindling, leaving her with burns over nearly half her body. She did not know that eighteen days later, she would be gone.

She should have known. Not because it was her job to know, but because it was someone else's job to teach her -- to protect her. Someone else's job to say "no" when the risks were greater than her preparation. Sheri did not fail the system. The system failed her.

This is the fire you can't see. The fire that starts not with flame, but with complacency. With silence. It begins with the checked box that says "training completed" when it wasn't, with the lab coat never worn, with the PI who assumes everyone knows what to do, but does not verify it, with the institution that forgets that real safety is not accomplished with just paperwork and posters.

In the years after Sheri's death, the University made promises. They formed committees. They improved protocols. They paid settlements. But the data tells a darker truth. In the eight years before Sheri's accident, there were 129 serious laboratory incidents. In the eight years after, there were 132. Sheri's death didn't mark a new laboratory safety initiative. It marked more missed opportunity.

We wanted to believe her death changed things. We still want to believe that. But believing doesn't make it so. Change is not something that automatically happens in response to a tragedy – it's something that must be fought for, day after day, even when no one is watching. Had Sheri only been given more information and emergency response training, she might be here to tell her own story. But we failed to protect her by not giving her information about the hazards and risks. This lack of information cost her life. We are responsible and we should never let it happen again.

At St. Andrews, no one died. But don't mistake that for a happy ending. The fire there displaced over 125 researchers. Years of work were destroyed. Lives were upended. Projects were derailed. Dreams were delayed. The toll was immense, just measured in a different currency. And still, the fire teaches us the same truth: safety is fragile, and its absence is brutal.

In both stories, what emerges is a chilling pattern: not recklessness, but routine; not malice, but misunderstanding; not sabotage, but silence.

Let's stop calling it "laboratory safety." That phrase has lost its power. It sounds passive, like a condition to be met. What we're really talking about is **accident prevention** – the active, vigilant, unrelenting work of stopping something before it starts. As one analysis put it, "Let's stop talking about lab safety. Let's start preventing accidents. Lives depend on it."

Accident prevention is not just goggles and gloves. It's culture. It's accountability. It's leadership. It's the courage to say, "We are not ready." It's the wisdom to recognize that the most dangerous fire is the one you don't see coming.

There are people alive today because someone took a moment to say "wait"— to double-check a label, to enforce a rule, to insist on a second pair of eyes. And there are people not here because no one did.

You might think a book like this is about tragedy. It isn't. It's about choice.

The choice every institution makes between investing in training or hoping luck will continue. The choice every professor makes between assuming a student understands or verifying that they do. The choice every laboratory makes between checking off a requirement or truly preparing people for the hazards ahead.

This is not just a book about Sheri Sangji or St. Andrews. It's about your laboratory, your workplace, your team. If you've ever walked into a laboratory, you've walked into a space where lives can be saved or lost, depending on what happens — or doesn't happen — before the work begins.

And that's the final lesson: **every accident is preventable** — but only if we choose to prevent it.

If you're a PI, a student, a facilities director, or an EH&S professional, you are holding a match. Will you light the way forward? Or will you, by inaction, light the fuse?

The fire you can't see is still smoldering. But there is time to extinguish it— if we act. If we lead. If we finally learn the lessons we should have learned, but didn't.

References

Basbug, G., Cavicchi, A., & Silbey, S. S. (2022). *Rank Has Its Privileges: Explaining Why Laboratory Safety Is a Persistent Challenge.* Journal of Business Ethics, 184(3), 571-587. https://doi.org/10.1007/s10551-022-05169-z

California Division of Occupational Safety and Health. (2009, December 23). *Investigation report: University of California Los Angeles (Case No. S1110-003-09).* Department of Industrial Relations, Bureau of Investigations.

California Division of Occupational Safety and Health. (2009, May 4). *Citation and notification of penalty: Inspection No. 312554796.* https://www.dir.ca.gov/

Chemical Safety and Hazard Investigation Board. (n.d.). *CSB laboratory incident data.* https://www.csb.gov/

Crase, M.M., "Underlying Factors of Laboratory Safety Behaviors in College Students Using the Theory of Planned Behavior and Risk Perceptions" (2022). *Graduate Research Theses & Dissertations.* 6951. https://huskiecommons.lib.niu.edu/allgraduate-thesesdissertations/6951

Gamble, J. (2011, April). Death in the lab. *Discover Magazine.* https://www.discovermagazine.com/technology/death-in-the-lab

Guyton GP. A brief history of workers' compensation. Iowa Orthop J. 1999;19:106-10. PMID: 10847524; PMCID: PMC1888620.

Henderson, T. (2025, March 27). The cost of lab safety: Why it's worth every penny. *Lab Manager.* https://www.labmanager.com/the-cost-of-safety-18468

Kemsley, J. (2019, January). 10 years after Sheri Sangji's death, are academic labs any safer? *Chemical & Engineering News*. https://cen.acs.org/safety/lab-safety/10-years-Sheri-Sangjis-death/97/i3

Lemonick, S. (2020, November 1). Who pays when a graduate student gets hurt? *Chemical and Engineering News*. https://cen.acs.org/safety/lab-safety/pays-graduate-student-hurt/98/i42

Lovell, B. (2018, September 4). Can a Woman Who Lost Her Arm In A Lab Explosion Sue UH? *Honolulu Civil Beat*. https://www.civilbeat.org/2018/09/can-a-woman-who-lost-her-arm-in-a-lab-explosion-sue-uh/

Notman, N. (2023, August 7 - updated 2023, Sept. 25). How to prepare for a lab-tastrophy. Chemistry World. chemistryworld.com/features/how-to-prepare-for-a-lab-castrophy/4017740.article

Occupational Safety and Health Administration. OSHA Memo, January 7, 2025. OSHA Penalties. https://www.osha.gov/penalties

Standler, R.B. (2013, March 18). *Injuries in School/College Laboratories in the USA*. www.rbs2.com/labinj2.pdf

Weizman, H. (2011). *Knowledge of the 2008 UCLA incident* [Conference presentation]. American Chemical Society, Division of Chemical Health and Safety Annual Meeting.

Sheri Sangji / UCLA Laboratory Fire

1. Allen, K. (2014, March 30). A young lab worker, a professor and a deadly accident. Toronto Star. Riskmanagement.nd.edu/assets/444869/ucla_lab_accident. pdf

2. Benderly, B. L. (2014, July 9). The landmark Patrick Harran case ends inconclusively. *Science.* https://www.science.org/content/article/landmark-patrick-harran-case-ends-inconclusivelyWikipedia+1Wikipedia+1

3. Chawla, D. S. (2018, September 6). Charges dropped against UCLA chemistry professor Patrick Harran for death of Sheri Sangji after lab fire. *Chemical & Engineering News.* https://cen.acs.org/safety/lab-safety/Charges-dropped-against-UCLA-chemistry/96/web/2018/09Chemical & Engineering News+2Wikipedia+2Wikipedia+2

4. Chawla, D. S. (2019, January 14). 10 years after Sheri Sangji's death, are academic labs any safer? *Chemical & Engineering News.* https://cen.acs.org/safety/lab-safety/10-years-Sheri-Sangjis-death/97/i3

5. Christensen, K. (2009, March 1). Deadly UCLA lab fire leaves haunting questions. *Los Angeles Times.* https://www.latimes.com/archives/la-xpm-2009-mar-01-me-ucla-fire1-story.htmlWikipedia+1Wikipedia+1

6. Christensen, K. (2012, January 21). Report faults professor, UCLA in death of lab assistant. *Los Angeles Times.* https://www.latimes.com/archives/la-xpm-2012-jan-21-la-me-ucla-lab-fire-20120121-story.htmlWikipedia+1Wikipedia+1

7. Christensen, K. (2014, June 20). UCLA chemistry professor avoids prison time in fatal lab fire case. *Los Angeles Times.* https://www.latimes.com/local/la-me-ucla-lab-fire-20140621-story.htmlWikipedia+1Wikipedia+1

8. Christensen, K. (2014, October 16). UCLA's legal fees in fatal lab fire case neared $4.5 million. *Los Angeles Times.* https://www.latimes.com/local/lanow/la-me-ln-ucla-legal-fees-lab-fire-20141016-story.htmlWikipedia
9. Stemwedel, J. D. (2012, January). Doing good science: The Sheri Sangji case. *Scientific American Blogs.* https://blogs.scientificamerican.com/doing-good-science/doing-good-science-the-sheri-sangji-case/janetstemwedel.com
10. Torrice, M. (2012, July 31). Patrick Harran and L.A. district attorney reach deferred prosecution deal in Sheri Sangji case. *Chemical & Engineering News.* https://cen.acs.org/articles/90/web/2012/07/Patrick-Harran-LA-District-Attorney.htmlWikipedia
11. Wikipedia contributors. (n.d.). Sheri Sangji case. *Wikipedia.* Retrieved [insert retrieval date], from https://en.wikipedia.org/wiki/Sheri_Sangji_caseWikipedia

University of St Andrews Laboratory Fire (2019)

1. Chawla, D. S. (2019, February 19). Fire that started from 'a routine post-experiment wash-up' closes building for a year. *Chemical & Engineering News.* https://cen.acs.org/safety/lab-safety/Fire-started-routine-post-experiment/97/web/2019/02Chemical & Engineering News+1news.st-andrews.ac.uk+1
2. McLaughlin, M. (2019, February 12). Blaze destroys years of medical research. *The Times.* thetimes.com/uk/Scotland/article/years-of-work-lost-in-st-andrews-blaze-Ombm5g987

3. University of St Andrews. (2019, February 15). Anatomy of a fire. *University of St Andrews News.* https://news.st-andrews.ac.uk/archive/anatomy-of-a-fire/news.st-andrews.ac.uk

4. University of St Andrews. (2023, May 2). Phoenix Rising from the Ashes' – world-leading biomedical lab re-opens after devastating fire. *University of St Andrews News.* https://news.st-andrews.ac.uk/archive/phoenix-rising-from-the-ashes-world-leading-biomedical-lab-re-opens-after-devastating-fire

5. Reddit user u/[username]. (2019, February 11). Chemistry building caught fire in St Andrew's [Online forum post]. *Reddit.* https://www.reddit.com/r/chemistry/comments/ap8n2p/chemistry_building_caught_fire_in_st_andrews/

Appendix A: Laboratory Safety Resources

This appendix provides a curated list of authoritative resources for those seeking to improve laboratory safety, enhance risk mitigation practices, and prevent incidents like the ones detailed in this book. These tools span regulatory agencies, academic institutions, professional organizations, and safety-focused publications.

🔒 Government and Regulatory Guidance

Occupational Safety and Health Administration (OSHA)
Laboratory Safety Guidance and Chemical Hygiene Plans
https://www.osha.gov/laboratory-safety

National Institute for Occupational Safety and Health (NIOSH)
Hierarchy of Controls for Risk Reduction
https://www.cdc.gov/niosh/topics/hierarchy

U.S. Chemical Safety Board (CSB)
Incident Reports, Safety Videos, and Lab Data
https://www.csb.gov

□ Professional Organizations and Training Platforms

American Chemical Society (ACS) – Safety Hub
Resources for lab safety education, best practices, and safety culture development
https://www.acs.org/safety

ACS – Safety in Academic Chemistry Labs (SACL)
Free guides for undergraduate and graduate-level safety education
https://www.acs.org/safety/education

Lab Safety Institute (LSI)
Workshops, audits, and safety program development
https://labsafetyinstitute.org

National Research Council – *Prudent Practices in the Laboratory*
The definitive lab safety manual from the National Academies
https://www.nap.edu/catalog/12654

🎓 University Resources and Case-Driven Insights

University of California Center for Laboratory Safety (UCCLS)
Founded in response to the Sheri Sangji incident
https://cls.ucla.edu

Harvard University – Laboratory Safety Program
Comprehensive policies, SOPs, and training tools
https://www.ehs.harvard.edu/programs/laboratory-safety

Stanford Environmental Health & Safety (EH&S)
Includes hazard assessments, incident reporting, and training modules
https://ehs.stanford.edu/topic/laboratory-safety

📰 Recommended Reading and Reports

Henderson, T. (2025). *The Cost of Lab Safety: Why It's Worth Every Penny.*
Lab Manager.
https://www.labmanager.com/the-cost-of-safety-18468

Chemical & Engineering News – Lab Safety Collection
News articles, case studies, and regulatory insights
https://cen.acs.org/safety

University Safety Culture Task Force Report (ACS)
Strategies for improving safety culture in academic laboratories
https://www.acs.org/content/dam/acsorg/about/governance/committees/chemicalsafety/safety-culture.pdf

---

□□ Incident Learning and Peer Networks

Learning from Incidents (LFI) – AIChE CCPS
Detailed reviews of chemical process safety incidents
https://www.aiche.org/ccps/tools/learning-incidents-lfi

DCHAS-L Safety Discussion List
Peer-to-peer network for chemical safety professionals
https://dchas.org/resources/join-dchas/

Additional Case-Specific Coverage

Sheri Sangji Case – C&EN Retrospective (2019)
https://cen.acs.org/safety/lab-safety/10-years-Sheri-Sangjis-death/97/i3

St Andrews Lab Fire – C&EN News Brief (2019)
https://cen.acs.org/safety/lab-safety/Fire-started-routine-post-experiment/97/web/2019/02

Join Our Community

Join the Movement:

Help Build a Safer Future for Every Laboratory

We invite you to become a founding member of the ***Laboratory Accident Prevention Council***—a community of professionals, educators, advocates, and researchers united by one shared mission: to prevent the next tragic laboratory accident before it happens.

This book has shown, in painful detail, that many accidents are not unforeseeable or freak occurrences. They are the consequence of overlooked warnings, ignored best practices, and cultures that value results over responsibility. Too often, institutional inertia delays meaningful reform until it's too late. The death of Sheri Sangji, the fire at St. Andrews, and hundreds of other preventable incidents make one thing painfully clear: **the cost of inaction is far too high**.

That's why we are forming the Laboratory Accident Prevention Council (LAPC)—a cross-disciplinary alliance dedicated to:

- Sharing case studies and lessons learned from past incidents
- Advocating for stronger, enforceable safety standards in educational and research institutions
- Developing training tools, resources, and safety frameworks tailored for modern laboratories
- Encouraging open communication and a "speak-up" culture where safety concerns are addressed – not ignored

We can't wait for change to come from the top down. It must start with us — from researchers in the laboratory to EH&S officers, from university leaders to industry partners. By joining this Council, you help create a unified voice demanding safer practices, better policies, and a future where no one pays a price for preventable harm.

Help shape that future. Join us. Speak up. Take action. Together, we can stop calling these tragedies "accidents" and start calling them what they really are: failures we refuse to repeat.

For more Information: **https://apc-labs.org/**

Free Fume Hood Stickers for Safety Training

If you would like some of these labels reach out to experts@fumehoodcertified.com

Get Hoodie Gear

https://fumehoodtesterscatalog.com/collections/hoodie-gear

Take advantage of Fume Hood Certifier's extensive resources!

fumehood.academy
Information regarding our in-person workshops in Phoenix. A great event for any experience level.

fumehoodcertified.com
Learn about all that FHC has to offer and how we may able to help you

tricolorasertesting.com
Home of the Tri-Color Visualizer system. Learn how we utilize a modified ASHRAE110 large volume test to challenge fume hood's dynamic containment abilities.

fumehoodonlinecourse.com
Chip's last book, "Laboratory Fume Hoods Explained" converted into a free online course

fumehoodperformance.com
The home for all your fume hood testing data for easy sharing, review and analysis. Works seamlessly with our Q200 and Face Velocity Profiler data capture software

fumehoodsafety.com
We host several introductory level videos here for students and employees that are new to fume hoods to get a understanding of safe user practices

fumehoodsoftware.com
Learn all about our Fume Hood Performance Tracker here, includes a walk-through video and a live demo.

fumehoodtesterscatalog.com
A one stop shop for everything you need to perform ASHRAE110 compliant fume hood testing, from smoke testing to Q200 factory calibrations, we have it all!

fumehoodtestingnetwork.com
Directory of industry professionals that utilize our Tri-Color testing system.

Safety is No Accident (Blog)
https://fumehoodtesterscatalog.com/blogs/lab-safety-digest

YouTube
www.youtube.com/@FumeHoodCertified

LinkedIn
www.linkedin.com/company/fume-hood-certified-llc

Facebook
www.facebook.com/fumehoodcertified

Laboratory Fume Hoods Explained
(Free Online Course)
www.fumehoodonlinecourse.com

Bulk Sales

Help Spread the word.

We offer discounts for multiple copies. Reach out to us at experts@fumehoodcertified.com

Do you have a Medical Portal?

Just like people use medical portals to monitor their health, laboratory fume hoods now have their own health tracker—**Fume Hood Performance Tracker**.

It's your digital portal to ensure every fume hood is operating safely and efficiently.

Fume Hood Certified is offering a **free trial** to help you get started. Discover how easy it is to track performance, flag issues before they become problems, and document safety like never before.

Don't wait for a safety incident to start monitoring your hoods—be proactive.

Email us today at **experts@fumehoodcertified.com** and start your free trial of the Fume Hood Performance Tracker.

HOOD
IDENTIFICATION
NUMBER

23-R24 Y44

SCAN ME

FHC FUME HOOD CERTIFIED

Your lab's safety deserves the same attention as your own health. Let us help you make that possible.

Made in the USA
Las Vegas, NV
02 September 2025

27311954R00056